Floods and reservoir safety: an engineering guide

Institution of Civil Engineers
London, 1978

Cover photograph by courtesy of North of Scotland Hydro-Electric Board: Pitlochry dam drum-gates spilling

WORKING PARTY ON FLOODS AND RESERVOIR SAFETY
Chairman: G. M. Binnie, FRS, MA, FICE
E. J. K. Chapman, BSc(Eng), FICE
C. L. Clarke, MA, FICE
M. F. Kennard, BSc(Eng), FICE
J. H. Lander, OBE, MA, FICE
J. W. Seddon, OBE, BSc, FICE
Technical Secretary: F. M. Law, BSc, MICE
Administrative Secretary: P. B. E. Thompson, BSc(Eng), FICE

Published by The Institution of Civil Engineers, Telford House, PO Box 101, 26-34 Old Street, London EC1P 1JH, and produced and distributed by Thomas Telford Ltd at the same address

First published 1978

ISBN: 0 7277 0033 2

© The Institution of Civil Engineers, 1978

All rights, including translation, reserved. Except for fair copying, no part of this publication may be reproduced, stored in a retrieval system, or transmitted in any form or by any means electronic, mechanical, photocopying, recording or otherwise, without the prior written permission of The Institution of Civil Engineers.

Printed by David Green (Printers) Ltd, Kettering, Northamptonshire

Preface

The Working Party on Floods and Reservoir Safety was appointed by the Council with the following terms of reference:

'To collaborate with the Natural Environment Research Council project on floods in the drafting of an engineering design manual or manuals to replace the 1933 report *Floods in relation to reservoir practice.*'

It met 22 times.

For the purpose of soliciting criticisms of its approach, a *Discussion paper on reservoir flood standards* was prepared and published by the Institution.

The recommendations contained in the 1933 interim report were stated to be 'a guide to water engineers when determining their own particular requirements. The subject does not lend itself to rigid treatment by means of precise rules and regulations.' It is now possible anywhere within the UK to estimate the magnitude of floods with a far greater degree of accuracy than before but, in spite of this great achievement by the NERC project team, it soon became apparent to the working party that it was still impossible to lay down fixed rules to meet all the ramifications of catchment conditions and types of dam that exist in the UK; the discussion paper, therefore, also took the form of an engineering guide.

The discussion paper was explained by Mr Frank M. Law at a gathering of engineers and hydrologists at the Institution of Civil Engineers on 22 April, 1975. The paper was debated at the Flood Studies Conference held by the Institution of Civil Engineers on 8 May, 1975, and a further discussion took place at the BNCOLD symposium on the Inspection, Operation and Improvement of Existing Dams held at the University of Newcastle on 24 September, 1975.

After consideration of the opinions expressed at these two helpful debates the standards were revised. To ensure the proposals would have the support of those members of the profession responsible for the inspection and safety of major dams, all Panel 1 engineers were sent a questionnaire embodying the revised standards, to which nearly all of them replied. Their comments and criticisms were considered in detail and the suggestions received led to considerable and useful amendments to the original draft.

The working party wishes to thank the Panel 1 engineers, and all who took part in the debates, for their contributions. It thanks also staff of the Meteorological Office and members of the Institute of Hydrology, in particular Dr J. V. Sutcliffe, for their valuable suggestions.

The working party recommends that after ten years' experience has been gained by panel engineers in using this guide, its contents should be reviewed and, if deemed necessary, it should be revised. However, it is stressed that an attempt has been made to keep the guide to matters of choice and principle rather than hydrologic technique. Thus it should not require amendment simply because of each new scientific advance. The most fundamental question remaining is whether an interrelationship exists (in terms of biased probability or physical limits) between the three primary factors of initial reservoir level, flood magnitude and wave size. The guide suggests a cautious combination of these elements in the belief that dam safety can be maintained only if all of them are considered. Despite refinements in technique it should be borne in mind that no flood prediction is more than an estimate.

Contents

1. Introduction... 1
    1.1. Background ... 1
    1.2. Scope of guide... 2
2. Reservoir flood protection standards ... 5
    2.1. General ... 5
    2.2. The main factors ... 5
    2.3. Recommended standards ... 7
    2.4. Special cases ... 11
3. Derivation of design flood ... 13
    3.1. Objective ... 13
    3.2. Steps in flood derivation ... 14
    3.3. Rapid calculation of flood peak inflow ... 14
4. Reservoir flood routing ... 17
    4.1. Objective ... 17
    4.2. Recommended stages in routing calculation ... 17
    4.3. Gated spillways ... 18
    4.4. Siphon spillways ... 19
    4.5. Auxiliary spillways ... 19
    4.6. Temporary upstream storage... 19
5. Wave surcharge and dam freeboard ... 21
    5.1. Scope ... 21
    5.2. Wind speed ... 21
    5.3. Fetch ... 21
    5.4. Wave height ... 21
    5.5. Wave run-up ... 22
    5.6. Wave surcharge allowance ... 22
6. Dam construction floods ... 23
    6.1. Risks over limited periods ... 23
    6.2. Diversion structures ... 24
Appendix 1. Rapid assessment of flood capacity and freeboard at existing dams ... 25
    A1.1. Purpose ... 25
    A1.2. Procedure ... 25
    A1.3. Example ... 27
Appendix 2. Particular considerations when using the *Flood studies report* for reservoir safety ... 31
    A2.1. Introduction... 31
    A2.2. Factors for consideration ... 31
Glossary ... 37

References	... ...	41
Bibliography	... ...	45
Fig. 1.	Least total cost analysis	47
Fig. 2.	Dam failure flood flow v. dam height	48
Fig. 3.	Average hourly annual maximum wind speed ...	49
Fig. 4.	Once in ten year hourly maximum wind speed ...	50
Fig. 5.	RSMD: England and Wales	51
Fig. 6.	RSMD: Scotland	52
Fig. 7.	RSMD: Ireland	53
Fig. 8.	Flood peak intensity for impermeable reservoired catchments	54
Fig. 9.	Relationship between effective fetch, wind speed and significant wave height	55
Fig. 10.	Ratio of run-up height to design wave height	56
Fig. 11.	Typical smooth dam facing: maximum run-up ...	57
Fig. 12.	Rough open-jointed pitching: intermediate run-up	57
Fig. 13.	Thick rip-rap facing: minimum run-up	57
Fig. 14.	Flood routing graph	58

Tables

Table 1.	Reservoir flood and wave standards by dam category	6
Table 2.	Design wave heights	22
Table 3.	Seasonal variation in Probable Maximum Precipitation	32

INTRODUCTION

be followed in references[4,5] which cover the consultative process that the working party chose to adopt. It also had the benefit of the replies of most Panel 1 engineers to a confidential questionnaire about flood standards early in 1976. This guide should be read in conjunction with the *Guide to the flood studies report*[6] and its associated maps, which can be purchased separately.

In the following pages the working party has defined standards on a sufficiently broad basis to cover all major points of principle. It has also attempted to give sufficient guidance to avoid excessive discrepancies in the conclusions reached by different engineers when reviewing the safety of an existing dam. The recommendations made here are not in any way mandatory and it is recognized that there will always be the occasional dam that has to be treated as an exceptional case. However, the working party suggests that where the engineer feels it is right to depart from its recommendations the fact should be recorded in his inspection report.

2. Reservoir flood protection standards

2.1. General

Protection standards must resolve acceptably the conflicting claims of safety and economy. Although it is now considered possible to design a spillway for the total protection of a dam against overtopping, there is the clear possibility that a smaller spillway built at less expense would survive several generations without any disaster or damage occurring. However, it is not simply a matter of economic judgement. As the Institution's 1973 statement on social responsibilities states, the civil engineer '. . . should recognise the many factors which may defy expression in direct money values, particularly those which arise from effects on a community's way of life.'

It has become necessary to categorize dams in terms of the risk they may pose to life and property downstream. Some dams, even if overtopped, are most unlikely to breach; increasing experience of erosion-resistance makes it possible to introduce this additional factor into present-day standards.

2.2. The main factors

A crucial question when considering flood protection is the combination of circumstances that may arise in progressively rarer events. Three main factors have to be defined:

(a) initial reservoir level;
(b) flood inflow;
(c) concurrent wind speed.

Despite continually improving techniques for defining flood hydrographs, wave run-up and flood routing, there is no indication that the engineer can do other than make separately reasoned assumptions on the levels at which the three factors listed above should be set. A consultative process has been followed in order to ensure that the values recommended for these assumptions in this guide are generally acceptable as 'standards' within the profession. Although the general framework may have relevance abroad, the values published in the accompanying table refer strictly to British dams.

Initial reservoir level. When investigating reservoir safety, it is necessary to consider at what level the stored water could be when flood inflow commences. Given techniques which assess sufficiently long flood durations relative to the flood storage capability of a dam,

Table 1 Reservoir flood and wave standards by dam category

Category	Initial reservoir condition	Dam design flood inflow			Concurrent wind speed and minimum wave surcharge allowance
		General standard	Minimum standard if rare overtopping is tolerable	Alternative standard if economic study is warranted	
A. Reservoirs where a breach will endanger lives in a community	Spilling long term average daily inflow	Probable Maximum Flood (PMF)	0.5 PMF or 10 000 year flood (take larger)	Not applicable	Winter: maximum hourly wind once in 10 years (Fig. 4)
B. Reservoirs where a breach (i) may endanger lives not in a community (ii) will result in extensive damage	Just full (i.e., no spill)	0.5 PMF or 10 000 year flood (take larger)	0.3 PMF or 1000 year flood (take larger)	Flood with probability that minimizes spillway plus damage costs (Fig. 1); inflow not to be less than minimum standard but may exceed general standard	Summer: average annual maximum hourly wind (Fig. 3) Wave surcharge allowance not less than 0.6 m
C. Reservoirs where a breach will pose negligible risk to life and cause limited damage	Just full (i.e., no spill)	0.3 PMF or 1000 year flood (take larger)	0.2 PMF or 150 year flood (take larger)		Average annual maximum hourly wind (Fig. 3) Wave surcharge allowance not less than 0.4 m
D. Special cases where no loss of life can be foreseen as a result of a breach and very limited additional flood damage will be caused	Spilling long term average daily inflow	0.2 PMF or 150 year flood	Not applicable	Not applicable	Average annual maximum hourly wind (Fig. 3) Wave surcharge allowance not less than 0.3 m

Notes
Where reservoir control procedure requires, and discharge capacities permit, operation at or below specified levels defined throughout the year, these may be adopted providing they are specified in the certificates or reports for the dam.
Where a proportion of PMF is specified it is intended that the PMF hydrograph should be computed and then all ordinates be multiplied by 0.5, 0.3 or 0.2 as indicated.

the working party considers that it is preferable to commence any routing calculation from a stable situation prior to a flood sequence. The alternative of postulating an antecedent flood that creates a starting water level has distinct drawbacks.

Flood inflow. It is necessary to specify a design flood, in combination with wave action, which the dam must be capable of withstanding. The passage of this flood through the reservoir should cause no fundamental structural damage to the dam. However, it is normally uneconomic to provide a waterway below the dam that is sufficiently large to contain the dam design flood outflow within its banks. Damage associated with rare overbank flows below or alongside spillway stilling basins may well be tolerated without risk to the integrity of the dam. Similarly, there are situations where it is not essential that the spillway channel should be hydraulically designed to carry all of the dam flood outflow. Some engineers prefer a clear distinction between the dam design flood and the spillway design flood but, as by definition only the former is vital to reservoir safety, this guide concentrates on the standards required for the bigger flood.

Concurrent wind speed. In the British climate, major rainfalls of several hours' duration can result from very different storm systems. Intense depressions can be expected to have relatively steady high winds over large areas, whereas thunderstorm clusters can exhibit very different wind patterns within a few kilometres. Sometimes storms may become stationary over an area, with low level air rushing in; other storm cells can move in a line, possibly up or down a valley, with consequent wind-veering. Waves generated on lakes during a storm, and more particularly at the time of peak flood surcharge at the dam, vary widely as a result. Records of past extremes suggest that a wave allowance should always be made and that it should be a more cautious one where there is a potential risk to human life. This allowance will provide a margin of safety in many (but not all) rare floods.

2.3. Recommended standards

In Table 1 are set out the standards which are appropriate for the wide variety and scale of dams covered by British safety legislation. To apply them it is necessary to route the appropriate dam design flood inflow using the corresponding initial reservoir condition and to obtain two levels, one being the theoretical flood surcharge level and the other being the total surcharge level; the latter includes the appropriate allowance for wave run-up caused by the wind speed given in Table 1 (or the minimum wave surcharge if that is greater), this wave surcharge allowance being sufficient to prevent overtopping reaching quantities that would hazard a dam crest.

For fill dams, the elevation of the crest of the dam will be governed by one of two conditions, the first being that the flood surcharge should not exceed the dam crest level, normally the crest roadway level; if the flood peak is particularly prolonged the flood surcharge level may have to be lower still to avoid harmful leakage through the road foundations above the dam core. The second condition is that the total surcharge must not overtop the dam wave wall. If there is no wave wall, the dam crest level has to be high enough to contain the total surcharge. On existing old dams with wave walls, it is recommended that caution should be exercised in taking wave walls into account, a frequent weakness being the presence of access gaps; in some cases it may be advisable either to ignore them in considering the safety of the dam or to reconstruct them. In some cases, where Table 1 produces much higher flood estimates at fill dams than were previously anticipated, it will be necessary to check the stability of the dam under the new estimated conditions.

For concrete or masonry dams designed over most of their crest lengths as spillway dams the concept of freeboard is irrelevant, and instead it is necessary to be certain that the flood loading derived from Table 1 does not go beyond structural design limits.

It may not always be necessary to provide for the general standards applicable to each category. If overtopping by events up to the Probable Maximum Flood (PMF) would have a negligible (say 1%) chance of breaching the dam, a minimum standard can be adopted. This is particularly the case with concrete and masonry dams where the additional head would not be expected to upset their structural balances. Where Table 1 refers to overtopping it means overtopping due to excessive still-water level during the routed flood and not to wave 'slop'. Whereas the minimum standard can be justified at an existing dam in a situation which is well understood, it would not be appropriate for the design of a new dam.

Although Table 1 may appear complex at first sight, it is designed to take account of those factors which are weighed together by panel engineers during dam inspections. Its main intentions are to ensure that, where a community could be endangered by a dam, the risk of any failure caused by a flood is virtually eliminated, but in other cases to keep expenditure to a scale justified by the risk. The justification of the initial reservoir level and wind speed criteria is covered in Chapters 4 and 5 and is not duplicated here. However, attention is drawn to the minimum wave surcharge allowance which varies by dam category. This is justified partly by the experience gained during the last 45 years since the 1933 interim report was published and partly by the need to allow for the vagaries of reflected waves and wave pluming at dams.

Category A dams. It is considered that public opinion will not accept conscious design for a specific threat to a community, even though it tolerates to an extent both random and accidental loss of life. Consequently, no dam above a village or town should be designed knowingly with a definite chance of a disastrous breach due to the under-provision of spillway capacity. A community defies definition in a few words; it is considered that inspection of any valley will soon reveal whether the presence of a hamlet, school or other social group means that a dam at its head should be in category A. Road and rail traffic caught in a valley flood would only accidentally be involved and would not by itself justify category A. A more difficult situation exists where an occasional camp site exists in the holiday season alongside a reservoired river; if, for example, this is in regular use by school parties it could well justify a community rating, but if it is frequented by a few unrelated short-stay individuals it need not.

Category B dams. Category B(i) is intended to refer to inhabitants of isolated houses and, for example, to treatment plant operators in a works immediately below a dam. (These situations lend themselves to taking measures to buy out the property or to arrange flood escape routes where appropriate.) Category B(ii) refers to extensive damage, including erosion of agricultural soils and the severing of main road or rail communications.

Category C dams. Category C covers situations with negligible risk to human life and so includes flood-threatened areas that are 'inhabited' only spasmodically; e.g., footpaths across the flood plain and playing fields. In addition this category also covers loss of livestock and crops.

Category D dams. Many small reservoirs with low earth dams may cause no real problem, except that of replacement, if they wash out. These special cases, many of which are ornamental lakes kept full for aesthetic reasons, are given a separate category. A flood intense enough to cause failure of a dam would create some damage even if the valley was still in its natural state; the additional damage caused by the release of stored water may well be insignificant if the lake is small. So where the amount stored would add no more than 10% to the volume or peak of the flood it is recommended that the spillway need not pass more than the outflow from the 150 year flood (or 0·2 PMF if that is calculated more readily). The point of reference for calculating whether the dam is significant or not can be taken as the first site below the dam at which some feature of value exists (e.g., a mill or road bridge). The 1000 year flood hydrograph applicable to that catchment prior to dam construction can be used for making this 10% sensitivity test.

Dam break wave. Assessment of the physical effect of a potential dam failure and the consequent flood wave is far from straightforward. However, Fig. 2 indicates that the flow immediately downstream is influenced mainly by dam height.[7] Differences between dam types are not strongly marked where storage is substantial. Extrapolation of the curve for a lower dam demands caution because individual circumstances at the site are likely to be more noticeable. Computer programs[8] are becoming available to estimate where flood levels may reach as the wave passes downstream. Either depth or velocity of flow may pose a threat. Although results cannot be precise, such a calculation can help specify the risk posed by the dam, and hence its category.

Economic considerations. Some reservoirs pose no threat to life but their loss would have severe economic consequences. Providing that all the losses caused by a failure can be met by remedial works and compensation payments, the sizing of the spillway and freeboard is a matter of locating the economic optimum. A point can be obtained on the total cost curve (Fig. 1) by summing the expected damage costs associated with a spillway sized for a flood of a particular return period and the capital cost of providing that spillway capacity. By examining designs for various flood magnitudes between 150 years and PMF, the solution giving least total cost can be identified. The method is that which was commended to US engineers by the Task Committee on the Reevaluation of the Adequacy of Spillways of Existing Dams.[9] Their report includes an excellent, albeit fictional, worked example.

Provision is made in Table 1 for the use of an economic standard as an alternative. The strength of the least-cost method is its ability to reduce the arbitrary choice of standards which may have costly implications. However, the most economic solution over the long term may not be one that the owner can finance in the short term. Indeed the economic study itself may be expensive (although this need not always be so[10]). The economics of the situation can be self-evident when, for example, a water treatment works is sited immediately below a dam and the loss of its output would have grave economic consequences for industrial consumers. Even for those cases where the failure of a new dam would not pose a serious threat to existing property, the additional cost of providing protection against the Probable Maximum Flood may be relatively small and it may be prudent to do so in order not to limit future development below the dam. After an economic study the panel engineer should be free to adopt safer flood control works than the nominal minimum solution if his appreciation of the extra costs of greater protection so indicates. Table 1 contains an important qualification that the alternative economic standard should not be allowed to produce a result that involves more risk of overtopping than the minimum standard.

Rapid assessment of existing dams. Some situations demand a rapid assessment of the ability of an existing dam to cope with the combination of flood, waves and initial reservoir level that has been discussed earlier in this chapter. Appendix 1 sets out such a method. Although it cannot be precise, it does provide a synthesis of the most recent practical work in this field and should reveal consistently those dams which merit fuller investigation of their safety in times of flood.

Failure to meet recommended standards. Immediate action is required where dam freeboard is inadequate to contain the flood surcharge of the appropriate standard; this normally takes the form of a temporary lowering of top water level prior to remedial works being carried out. However, if a dam is found to be adequate to contain floods but unable to cope with the associated waves required by Table 1, the action to be taken will depend on local circumstances. Remedial measures such as a wall or rip-rap protection may be feasible. In some cases wave observation may prove that the dam is noticeably sheltered and enable the inspecting engineer to relax the normal wave standard. Note should be taken at earth dams of the extent to which spray is ever seen to pass over the crest on to the downstream slope; where this is significant, standards should not be relaxed.

2.4. Special cases

Gated spillways. Where a gated spillway is employed, high standards of maintenance are required and regular operation is essential. This is particularly so where gates form the sole outlet for flood waters and where the dam is of the fill type. The provision of gates cannot be encouraged unless the dam owner's organization is sufficiently large to ensure the necessary maintenance and operation even under emergency conditions. Because the chance of human error or mechanical failure cannot be discounted, it is considered that the following additional safeguards should be provided at dams where overtopping would be likely to cause a breach:

(a) a minimum of two gates;
(b) the ability of the remaining gates to pass at least the 150 year flood if one gate is out of action;
(c) if the dam falls in category A, automatic operating equipment with the necessary standbys and option for manual override.

Minimum spillway capacity. A few reservoirs are so large relative to their direct catchment that routed flood outflow is very small. In such cases the spillway should have not less than a minimum capacity and be constructed so that it will not tend to block with debris or ice. The capacity, at a head which leaves a freeboard equal to the wave surcharge allowance, should be not less than

$$\frac{\text{RSMD}}{20} \times 0.15 \text{ m}^3/\text{s per km}^2 \text{ of catchment area}$$

RSMD is the *Flood studies report*'s index for flood-producing rainfall; it is mapped in Figs 5–7. The equation is equivalent to a daily runoff of 13–65 mm, depending on location. It should be noted that snowmelt can be a controlling factor in some circumstances; the *Flood studies report*'s recommendation (Volume I §6.8.3) is for a melt rate of 42 mm/day over the catchment, treated as additional precipitation until all the snow has cleared.

Reservoirs without direct catchments. There are open service storages and pumped storage reservoirs with no natural catchment. For these no spillway is recommended, provided

(a) adequate fail-safe provision exists to ensure that the piped inflow cannot continue to the point of overtopping the embankment;

(b) the total freeboard above normal top water level is adequate to contain run-up resulting from maximum wind conditions. (This may be calculated using the wind/wave column of Table 1, testing the appropriate category against values 50% higher than those shown; where this approximate approach has important financial implications the Meteorological Office should be contacted for wind speed estimates of appropriate rarity (i.e., category D—150 year, category C—1000 year etc.)).

Grass-covered embankments and spillways. Earth embankments with a level crest and a surface roadway may well resist some overtopping successfully. CIRIA[11] suggest that well chosen grass on an embankment can withstand the following velocities:

(a) up to 2 m/s for prolonged periods (say more than 10 h);
(b) 3–4 m/s for several hours;
(c) up to 5 m/s for brief periods (say less than 2 h).

This research opens the way to the better design of auxiliary grass-covered spillways to take occasional flood flows.

Fill dams in deep valleys. When fill dams are sited in deep valleys there are substantial catchment areas which in a severe storm drain towards one or both flanks of the dam and this can result in considerable damage to the embankment. In addition to drains at the junction of the downstream slope with the valley sides, additional drainage measures (such as catchwaters) may be necessary to direct storm water clear of the embankment.

Discharge alterations. Where alterations are about to be made to an uncontrolled spillway or to the operational rules governing flood discharge gates, the undertaking concerned should draw the facts to the attention of owners of lakes that lie lower down the river; the authority responsible for flood plain management should also be notified.

3. Derivation of design flood

3.1. Objective

To apply the suggested flood standards it is necessary to determine floods of magnitudes between the 150 year event and the physical upper limit. It is not sufficient simply to specify any of these design floods by the magnitude of the peak, because the shape and volume of the associated hydrograph also determine the extent to which it will be modified by passing through reservoir storage. The appropriate method for obtaining a complete design hydrograph is by describing the catchment response to a design storm using a unit hydrograph; only where storage is so small that flood attenuation is negligible do other methods of peak flow estimation become acceptable.

The working party has interpreted its terms of reference to mean that this document need not reproduce standard technical procedures published elsewhere; the reader making comprehensive flood calculations should refer in the first instance to the *Guide to the flood studies report*.[6] It contains worked examples embodying unit hydrographs whether for gauged or ungauged catchments. This chapter has the complementary objective of

(a) identifying the main steps in any unit hydrograph method, whether precisely following the *Flood studies report* approach or otherwise;
(b) indicating the differences between estimating a flood of given probability or the Probable Maximum Flood by the *Flood studies report* form of this method;
(c) introducing a rapid method of calculation which may be used at existing dams as a first estimate of flood potential.

In Appendix 2 is a discussion of details in the *Flood studies report* method which arise when it is applied to reservoired catchments.

The difficulty in flood estimation comes from the multiplicity of natural factors, each with its own probability distribution, which combine to produce floods. Scope exists for improving most techniques and such developments are to be encouraged. Nevertheless, recent years have seen substantial progress in the explanation of processes occurring during rare floods and in the estimation of the range of future flood potential.

3.2. Steps in flood derivation

The steps in flood derivation can be summarized as follows:
   (a) deduce appropriate design storm duration;
   (b) compute catchment rainfall of that duration and of design severity;
   (c) distribute design rainfall according to an appropriate profile of intensity;
   (d) apply storm losses due to detention and infiltration;
   (e) multiply the increments of remaining rainfall excess by the unit hydrograph ordinates and add to produce the storm runoff;
   (f) add baseflow to obtain the complete flood hydrograph.

Each of these steps is carefully defined in the *Guide to the flood studies report*, and formulae, maps or graphs are provided for each of them. It should be noted that the *Flood studies report* method is an integrated whole designed primarily to ensure that a flood peak of required rarity is obtained. By simulating this design sequence its authors were able to specify the rainfall probability to produce a desired runoff probability by a single graph, but this depends implicitly on their calculation assumptions being followed. However, they vary these assumptions for the estimation of the Probable Maximum Flood by

   (a) adopting a different profile of storm intensity with time;
   (b) adding snowmelt at a uniform rate to precipitation (up to a mapped total);
   (c) reducing the time to peak of the unit hydrograph by one third and so raising its peak flow rate by 50% (this change is based on observations of severe events instead of using the mean time to peak of a wider range of floods).

It is desirable that graphs and equations based on mapped characteristics of catchments should not be used until there has been a visit to the site concerned. For a thorough appraisal the steps (d)–(f) in the first paragraph of this section should be carried out not only by using such equations but also separately on the basis of any local flow data that exist. In the absence of near agreement between the two approaches, the more pessimistic for spillway sizing should be adopted unless definite reasons exist for preferring one over another.

Alternative flood figures could be avoided only if a rigorous and inflexible code of practice were to be laid down. Inevitably such a code would be unable to deal with the great variety of catchment conditions and dam types that exist in the UK. It must be left to the engineer responsible to produce a reasoned flood estimate which satisfactorily reconciles recognized methods with local data.

3.3. Rapid calculation of flood peak inflow

Figure 8 summarizes on a single graph the range of flood peak intensity that may be expected from impermeable rural catchments. It

DERIVATION OF DESIGN FLOOD

gives an estimate of the Probable Maximum Flood of an undulating catchment by a *Flood studies report* formula. Adjustment factors for different terrains or for less rare floods are given. Although the graph is part of the rapid method of checking flood rise and freeboard at existing dams, it is also possible to use it to check approximately whether unit hydrograph computations of the type described above have produced results at or near the usual intensity for the area concerned. However, because it is based on the most impermeable category of the five soil classes mapped in the *Flood studies report*, it may give estimates that are too high if it is applied in the other four classes of area.

4. Reservoir flood routing

4.1. Objective

The calculation of the change that takes place in the hydrograph of a flood as it passes through a reservoir is a standard procedure. The purpose of this chapter is to comment on the decisions and assumptions that are required at the outset of a routing calculation or when reviewing initial results. Such routing calculations produce estimates for

(a) flood level attained in the reservoir;
(b) time lag between inflow and outflow peaks;
(c) the spillway flow hydrograph resulting from chosen inflow.

Computations may be performed manually[12] or by computer.[13] Where valid approximations can be made, graphical solutions shorten routing computations. The most recent of these[14] has been developed specifically to assist with the rapid assessment of floods at existing dams (Appendix 1). However, full calculations should be adopted when either a new or an enlarged spillway is to be constructed. In some cases routing may be immaterial due to the small reservoir storage above normal top water level compared with the expected flood magnitude; it is easy to check whether this will be so by reference to the attenuation graph of Appendix 1 (Fig. 14).

4.2. Recommended stages in routing calculation

Recommendations covering choices in routing estimates are summarized below and justified subsequently.

(a) *Initial reservoir level and spill rate.* These are given in Table 1; where needed, compute mean inflow from the site runoff record or catchment rainfall less evaporation; for hydroelectric reservoirs assume generators are operative.

(b) *Draw-off and releases during floods.* Ignore draw-off and releases; for hydroelectric reservoirs assume generators are unable to operate.

(c) *Level/discharge relationship.* Incorporate any restriction on spillway flow due to downstream culvert, bridge, channel, dropshaft or tunnel capacities and record the relationship up to the level at which overtopping would begin (where relevant); at higher levels, ratings with and without the overtopped section blocked off are normally required (the special cases of gated, siphon and auxiliary spillways are discussed later).

(d) *Level/capacity relationship.* Compute from graph of lake elevation against area flooded.

(e) *Reservoirs in series.* Route through the upper reservoir, add the inflow of the catchment downstream and route through the next highest reservoir and so on, preserving the timing of events dictated by a single design storm (Appendix 2).

Initial reservoir level conditions have to be specified that take proper account of antecedent flow from a preceding flood and of the control on water level that the function of the reservoir exerts. Table 1 therefore adopts the safe assumption for category A dams that they will be full and spilling prior to the design storm and makes the same assumption for category D lakes because typically these are kept full as amenity features. As intermediate-category dams are predominantly for water supply and experience continual draw-off they are not likely to be spilling prior to a rare flood and some relaxation from category A standards seems justified.

The problem of antecedent flow is acute only where design storm duration exceeds one day. For such large catchments, where insufficient is known about the spacing in time between storms, care must be taken to examine local records of long wet spells. The temptation to route only short intense floods has to be avoided.

Increasingly, reservoirs are being operated to achieve specified drawdown levels, varying through the year, that provide substantial flood control benefits. Where these levels can be achieved with a high degree of probability because of control procedures and discharge capacities, these can be adopted for routings. The dominant level for the season concerned (i.e., summer or winter) is suggested. Where the adequacy of the spillway or freeboard depends on this starting level it should be specified in the certificates and inspection reports for the dam.

If a flood routing calculation produces a very small outflow maintained for a considerable period there is a risk that an alternative and more critical sequence of flood-producing events has not been identified. For this reason Chapter 2 specifies a minimum spillway capacity as a safeguard against floods produced by wet periods of many days' duration.

4.3. Gated spillways

The use of gated spillways has been limited largely to Scotland. The long experience of the North of Scotland Hydro-Electric Board covers as severe a climatic regime as any other dam owner may anticipate. Their gate reliability is high due to careful maintenance; such work results in a typical flood gate being out of service for less than one week per decade and this can usually be timed to be in dry weather. As with few exceptions there is more than one gate to each dam, the others remain available to handle an extreme flood.

Gate-controlled hydro-power dams frequently have a complex operating regime[15] which is particularly important in basins subject to prolonged floods and containing several interconnected dams. This, and the recommendation that checks should be made on potential flood levels with one gate immobilized, can require extensive routing studies but these raise no new issue of principle.

Controlled discharges from some dams during floods can be significant in limiting the volume of water that will be impounded. In making allowance for this in routing estimation no reliance should be placed on being able to raise the discharge rate during a rare storm unless deliberate and effective contingency plans exist to ensure this. Site access and availability of power or personnel for gate or valve changes can all be affected by the storm concerned.

4.4. Siphon spillways

Siphon spillways have an unusual level/discharge relationship but now that their hydraulic characteristics are better understood and can be controlled by air regulation[16] it can be expected that their use will be more widespread. The need to limit the rate of rise of flood levels downstream of them to no more than the natural rates prior to dam construction will be readily understood; any discharge that could create a vertically-faced wave is particularly dangerous to other river users.

4.5. Auxiliary spillways

In some cases auxiliary spillways have been, or will have to be, constructed to carry the recommended flood flows that are necessary for modern safety standards. It is not possible to recommend that flows carried by auxiliary spillways should be a fixed proportion of the design flood or that they should only operate with a specified frequency. Experience shows that the nature of the site normally dictates the few available options, each of which can readily be tested by an appropriate routing calculation.

4.6. Temporary upstream storage

An increasing number of reservoired catchments are no longer natural but contain various constraints on the occurrence of intense flood runoff. Examples are substantial railway and motorway embankments through which reservoir tributary streams are culverted. Where these would temporarily impound significant quantities of water during a flood, and where they are unquestionably stable, they may be treated as additional reservoirs to be incorporated in the routing. Where an existing spillway would be inadequate but for the protection afforded by upstream storage, this fact should be recorded in the inspecting engineer's report in order to forewarn the supervising engineer should the storage be removed; the closure of railways and removal of their embankments is a case in point. It should be borne in mind that railway embankments, which are relatively narrow and unsurfaced, may breach in a design flood, in which case they can be more of a danger than a protection.

5. Wave surcharge and dam freeboard

5.1. Scope
In this chapter is outlined a method of estimating the wave surcharge allowance referred to in Table 1 for preventing overtopping from reaching quantities that could endanger a dam. Except on wind speeds, very little directly applicable research has been done on the various factors governing overtopping of dam crests by waves on small reservoirs. The validity of the results obtained by the method should therefore be considered as being sufficient only for the purpose of considering wave surcharge in relation to the other factors in Table 1, and not of general application.

5.2. Wind speed
Figures 3 and 4 are wind speed maps for the UK, produced by the Meteorological Office for the Institution using data available up to 1977. They correspond to the wind speed standards recommended in Table 1. If sufficient local wind speed data are available for probability analysis, they may be used to refine the diagram.

One can ascertain whether the design flood will occur in winter or summer on the basis of the seasonal variation in precipitation given in Appendix 2 (with the addition of snowmelt for winter PMF events) and with a knowledge of the design storm duration. Then in accordance with Table 1 the appropriate wind speed can be taken from Fig. 3 or Fig. 4 for the corresponding season. Even when the Probable Maximum Flood occurs in summer, for categories A and B higher wind speed in winter may combine with a smaller flood and become the factor governing maximum reservoir level; in such cases it is necessary to evaluate for both seasons.

5.3. Fetch
For use with the wind speeds specified in Table 1, it can be assumed that the effective reservoir length over which winds can build up waves (referred to as the fetch) is the average reach over a 90° arc across the reservoir to the dam.

5.4. Wave height
It is most useful to characterize the design wave height as that which is exceeded by only a stated percentage of the whole wave population.[17] The significant wave height H_s of a regular wave train has been defined as the average height of the highest one-third of all

Table 2 Design wave heights

Dam type	Crest	Downstream slope	Design wave height	Percentage of waves above H_D
Concrete/masonry	—	—	$0.75H_s$	33
Rock fill	Road	—	$1.0H_s$	13
Earth fill	Road	Selected[11]/reinforced grass	$1.1H_s$	9
Earth fill	Road	Random grass	$1.2H_s$	6
Earth fill	Grass	Random grass	$1.3H_s$	4

waves. This height has been shown to be equalled or exceeded by 13% of the waves generated by a given wind speed. Having obtained the wind speed and the fetch, the significant wave height H_s can be determined from Fig. 9.

The extent to which waves can be permitted to go over the crest of a dam during a dam design flood depends on the type of dam; a concrete dam, for example, is more resistant to erosion than an earth-fill dam. The allowable height, referred to as the design wave height H_D, can be more or less than the significant wave height H_s as the case may be. It is convenient to express H_D as a multiple of H_s. Recommended ratios of H_D to H_s are given in Table 2.

5.5. Wave run-up

Depending on its smoothness and permeability, the upstream slope of a dam also influences the height to which a wave will run up. Based on the upstream slope, Fig. 10 relates the ratio of run-up height to design wave height H_D for three different surfaces, smooth surface as Fig. 11, rough stone or shallow rubble as Fig. 12 and thick permeable rip-rap as Fig. 13. It is left to the engineer concerned to relate his dam to those illustrated and to interpolate on Fig. 10 accordingly.

5.6. Wave surcharge allowance

To summarize, the recommended procedure is to obtain the wind speed from Fig. 3 or Fig. 4, the fetch from a plan of the future or existing reservoir, the significant wave height from Fig. 9, the design wave height from Table 2, and the run-up height from Fig. 10. Except in those cases where the minimum value applies, the result is the minimum wave surcharge allowance for use in Table 1.

6. Dam construction floods

6.1. Risks over limited periods

Normally dam construction requires some temporary river diversion works to prevent or minimize damage by flooding before completion. The choice of a construction flood against which to design is a compromise between economy and safety. Normally it is accepted that a higher risk can be run during a limited construction period because of the expectation that any damage would be restricted to the site, particularly in the case of concrete structures which are less susceptible to breaching. Thus at many upland sites it has been the practice to adopt a fraction, say one half, of the Normal Maximum Flood as defined from the curve in the 1933 interim report. A construction flood of this magnitude is not particularly rare and many dam sites have suffered limited flood damage as a result. With modern plant, dams are built more rapidly, the period of risk is shortened and therefore the incidence of damage has probably decreased.

Where an arbitrary criterion is appropriate it should be acceptable to design for the flood that has only a 10% chance of being exceeded during the critical part of the diversion period.[19] Thus, if this period is one year, the ten year flood peak would be chosen. The percentage probability P_r that an event of recurrence interval T years or greater will occur at least once within the next r years can be taken as

$$\frac{P_r}{100} = 1 - \left\{1 - \frac{1}{T}\right\}^r$$

The only proviso is that flood peaks are presumed to be independent random events. Evaluating this formula shows, for instance, that the flood corresponding to 10% risk over a five year construction period has a 48 year return period. Simplifying, for practical purposes:

for 10% risk, the construction flood return period = duration of risk (years) × 10.

The duration of risk should be estimated pessimistically and should normally be taken as being at least one year.

Computing a flood of low return period is best done from flow records at the site concerned. Even using two or three years' data is likely to give a better estimate of flood frequency than a theoretical equation, provided the measurements are used to define the mean annual flood; this can then be multiplied up to estimate rare floods using the growth curves in the *Flood studies report*.

6.2. Diversion structures

Diversion proposals may well involve an upstream cofferdam that provides some storage at the inlet to a tunnel or culvert. However, it is not usual to compute a flood peak reduction due to routing; instead, the introduction of a small safety margin is preferred.

In some circumstances it is possible to design a structure to withstand erosive overtopping forces during construction and so avoid costly diversion works. This can be attractive at a narrow dam site and has frequently been adopted during the building of concrete dams. The armouring of a rock-fill dam near Canberra (Australia) successfully withstood a peak flood of 1.8 m over it in 1976; the specification had required that wire mesh protection should be welded into position at the cessation of each day's embanking.[20]

Appendix 1. Rapid assessment of flood capacity and freeboard at existing dams

A1.1. Purpose
Whilst comprehensive flood studies become more exacting, there is a continuing need for a rapid but safe method of assessing whether an existing dam can adequately contain floods and waves. Such a method can

   (a) indicate whether a dam has (or lacks) a large margin of safety;
   (b) help keep the balance between investigation costs and dam maintenance expenditure;
   (c) assist the inspecting engineer to make his initial appraisal while on site.

The 1933 report met this objective for a limited range of dams with graphs giving flood intensity per unit catchment area and flood reduction factors based on reservoir level rise and the proportion of the catchment inundated. Some rules were included to cover minimum required freeboard with and without a wave wall across the dam. This appendix attempts to solve this problem with fair accuracy for any UK reservoir (given only a map and the reservoir record book to assist a field appraisal).

It is intended that this appendix should be replaced in time with an improved version. The simplifying assumptions on which it is based may then be capable of refinement to improve its reliability. It is strongly recommended that the engineer should use this method only after he has 'calibrated' its probable accuracy and value at reservoirs where he has a thorough knowledge of flood potential. The theoretical background to the flood routing approximation has been published previously[14] and is not repeated here.

A1.2. Procedure
Knowledge of the dam and its setting will indicate which of the standards from Table 1 is appropriate. The natural flood peak to the dam, Q_i, unmodified by the reservoir, can be obtained from Fig. 8 once the local rainfall parameter RSMD has been taken from Figs 5–7 (this can be averaged by eye if it is a large catchment). The flood

flow should be multiplied as indicated to allow for alternative return periods and catchment terrain which is other than undulating. The graph implies that catchment soils are relatively impermeable and that urban development within the catchment is minimal.

To estimate the maximum potential of the reservoir for reduction of flood peak Q_i it is necessary firstly to assess the available freeboard. The amount required for wave surcharge can be found as follows.

(a) Assess the effective fetch F to the most exposed part of the dam. It is adequate to visualize from a map the average fetch length over a 90° arc; it will frequently approach (reservoir area)$^{1/2}$, but in thin and sinuous lakes it can fall to about one third of the single longest possible straight-line fetch.

(b) Read the corresponding significant wave height H_s from Fig. 9 at the wind speed criterion suggested in Table 1 and mapped in Fig. 3 or Fig. 4.

(c) Multiply H_s by the appropriate factor from Table 2 to obtain the design wave height H_D.

(d) Multiply H_D by the run-up factor shown in Fig. 10 to obtain design wave run-up on the upstream face concerned, this result being wave surcharge allowance.

Wave reflections, irregularities in wave distributions and pluming make precise calculation impossible. Notwithstanding the above calculation, no lesser value than the appropriate minimum freeboard allowance for wave run-up surcharge shown in Table 1 should be adopted.

For an existing dam lacking a wave wall, the permissible flood lift h_m from the initial reservoir condition (Table 1) is the height to the minimum dam crest level less the wave surcharge allowance computed previously. This can then be checked against the actual head h that will be reached on the existing weir as peak flood inflow is routed through reservoir storage. For the purposes of calculation, estimate reservoir area a at a head equal to 0.5 h_m (usually possible directly from the reservoir record book). Then proceed as follows.

(a) Calculate the head H on the spillweir (length B) that would correspond to flood peak Q_i if reservoir area was minimal; i.e.,

$$H = \left\{ \frac{Q_i}{CB} \right\}^{2/3}$$

An appropriate discharge coefficient C for high head situations has to be chosen for the weir.

(b) Calculate the storage ratio

$$S = \frac{aH}{Q_i T_p}$$

in consistent units (e.g., m^{2}m/(m^3/s s)). If T_p, the time to peak of the unit hydrograph, is not known it can be approximated for floods of given return period by

RAPID ASSESSMENT OF FLOOD CAPACITY AND FREEBOARD

$\left.\begin{array}{l}4600\ A^{1/4} \text{ in mountain areas} \\ 5300\ A^{1/4} \text{ in hilly areas} \\ 5900\ A^{1/4} \text{ in undulating areas} \\ 7900\ A^{1/4} \text{ in flat areas}\end{array}\right\}$ Multiply by 0.67 when T_p is used in PMF calculations (*Flood studies report*, Volume I §6.8.3b)

where A, the catchment area, is in square kilometres and T_p in seconds. (The relationship is simplified from *Flood studies report* equation I.6.15 for catchments of typical shape that are under 100 km² in area.)

(c) Using S in Fig. 14, find the proportion R (the attenuation ratio) of H that will actually occur in the given flood and so calculate the flood lift $h = RH$.

(d) If h (plus initial reservoir head where applicable) is greater than or closely approaches h_m, a full calculation by the best available methods is warranted.

If this calculation shows that the dam is capable of being overtopped, the assumption about permissible flood lift should be reviewed; in many cases it will be possible to show that the flood surcharge alone can be contained below crest roadway level and that, providing a crest wall exists which will deal with the wave surcharge, a safe situation exists.

A1.3. Example

To illustrate the above procedures an example is given below. It follows a descriptive format which can be condensed further once the calculation steps are appreciated.

1. Site   *Onemore Reservoir*

2. Grid reference

3. Large dam (by ICOLD definition)?   YES/~~NO~~

4. Topographic category of catchment: ~~mountainous~~/hilly/ ~~undulating/flat~~

5. Flood protection category of dam (Table 1):   A/B/~~C/D~~
   Principal reasons? (a) Community at risk
                    (b) Other risk to life
                    (c) Economic losses   ✓ *cascade effect on reservoir below*

6. Adopted flood standard (Table 1): general/~~minimum/financial~~
   Reasons for minimum standard?   *N/A (not applicable)*

7. Dam type: ~~rock fill/concrete and masonry~~/earth fill

8. Dam crest material: ~~grass~~/surfaced road
                        ~~other~~:

9. Downstream slope material: grass/~~reinforced grass~~
                             ~~other~~:

FLOODS AND RESERVOIR SAFETY

10. Upstream slope material (Figs 11–13):
    maximum run-up surface
    ~~intermediate run-up surface~~
    ~~minimum run-up surface~~
    Summary description *Regular blockwork*

11. Upstream slope   1:3

12. Dam crest level (without wave wall):
    nominal   189.05 m (Source: *Drg M 239/6*)
    measured low point 188.80 m (Date: *1.4.77*)
    measured typical point 189.10 m (Date: *1.4.77*)

13. Wave wall crest level: *N/A*
    nominal   m (Source:
    measured   m (Date:
    details of gaps etc.

14. Weir crest level: *Owner's*
    nominal   187.14 m (Source: *Drg 16.2.54*)
    measured   187.140 m (Date: *1.4.77*)

15. Weir crest length B:
    nominal   15.2 m (Source: *Record Book*)
    measured   15.24 m (Date: *1.4.77*)

16. Catchment area, km²: $A =$ *4.1*

17. Average annual rainfall, mm: $p =$ *1650*

18. Average flow, m³/s; categories A and D only:
    (average annual rainfall − loss) $\times A \times 3.2 \times 10^{-5}$ = *0.15*
        1650   *500*
    or as measured   = *N/A*

19. Net one day rainfall of five year return period, mm (Figs 5–7):
    RSMD = *57*

20. Peak inflow intensity, m³/s per km² (Fig. 8): $Q_m/A =$ *16.0*

21. Required standard flood Q_i as multiple of Q_m (Fig. 8 and item 6 above): 1.0/~~0.5/0.3/0.2~~

22. Factor, for terrain other than undulating, % (Fig. 8): *+5*

23. Standard flood [item 21 $\times (Q_m/A) \times A$] [1 + terrain factor/100]:
        1.0 × 16.0 × 4.1 × 1.05
    $Q = 68.9$ m³/s

24. Effective fetch, km: $F =$ *0.36*
    Method used: visual
        ~~Saville~~
        ~~other~~

25. Design wind speed, m/s (Table 1 and Fig. 3 or Fig. 4: for safety, for categories A and B adopt Fig. 4 only; i.e., no summer reduction in this rapid method): $U =$ *26.0*

RAPID ASSESSMENT OF FLOOD CAPACITY AND FREEBOARD

26. Significant wave height (approx), m ($UF^{1/2}/40$ or Fig. 9): $H_s = 0.39$
27. Wave height factor based on items 7–9 above (Table 2): 1.2
28. Run-up factor (Fig. 10; see also items 10 and 11 above): 1.7
29. Wave run-up, m ($H_s \times$ wave factor $\times$ run-up factor): 0.80
    $0.39 \times 1.2 \times 1.7$
30. Minimum freeboard allowance, m (Table 1): 0.6/~~0.4/0.3~~
31. Review wave surcharge, m (adopt larger of items 29 and 30):
    0.80
32. Spillweir discharge coefficient (metric) in $CBH^{1.5}$: $C = 1.86$
33. Initial reservoir level (Table 1), m {~~item 14 or~~ [item 14 + (item $18/CB)^{2/3}$]}: 187.17
34. Permissible flood lift (item 12 − item 33 − item 31): $h_m = 0.83$ m
35. Reservoir area at $0.5h_m$ above weir crest: $a = 158,300$ m^2
36. Unattenuated flood lift $(Q_i/CB)^{2/3}$ (this calculation can be made only if spillweir flow remains modular under flood conditions): $H = 1.81$ m
37. Time to peak of unit hydrograph: $T_p = 5050$ s
38. Storage ratio aH/Q_iT_p, m^{2}m/(m^3/s s): $S = 0.82$
39. Attenuation ratio (Fig. 14): $R = 0.86$
40. Flood surcharge (RH + item 33 − item 14): $h = 1.59$ m
41. *First check: ability of dam to provide adequate total freeboard*
    If without wave wall:
    Dam freeboard, m (item 12 − item 14): 1.66
    Flood and wave surcharges, m (item 40 + item 31):
    $1.59 + 0.80 = 2.39$
    If with wave wall: N/A
    Maximum dam freeboard, m (item 13 − item 14):
    Flood and wave surcharges, m (item 40 + item 31):
    (Whereas a subsequent full calculation gave
    $1.55 + 0.80 = 2.35$ m)

    CONCLUSION: Water ~~is not~~/is expected to overtop the dam in unacceptable quantities if anticipated waves occur

42. *Second check: ability of dam to retain flood water below dam crest level*
    Minimum dam freeboard, m (item 12 − item 14): 1.66
    Flood surcharge, m (item 40): 1.59

    CONCLUSION: Still water alone ~~is~~/is not expected to exceed the dam crest level

REMARKS:

(a) Consider wave wall or rip-rap.
(b) Run-up predictions will be less severe if item 25 above can be refined down to summer wind conditions.
(c) Items 34, 41 and 42 assume low points NOT made up to original nominal crest level but this is clearly necessary.

Appendix 2. Particular considerations when using the *Flood studies report* for reservoir safety

A2.1. Introduction

Chapter 6 of the *Flood studies report* covers the necessary calculations for estimating a flood hydrograph off any UK catchment which has less than 25% urban area and less than 33% of its area above any upstream lake. The *Guide to the flood studies report*[6] introduces the methods to be employed and gives examples of finding a flood of given probability, or the Probable Maximum Flood. In both cases the unit hydrograph technique is demonstrated. The only additional material required when using the guide is the *Flood studies report* maps volume (Volume V).

The *Flood studies report* covers floods in their widest sense and is not specifically directed towards floods in relation to reservoir safety. For this reason certain factors have to be taken into consideration when using it and these are described in the following paragraphs.

A2.2. Factors for consideration

Storm duration. The *Flood studies report* makes the important recommendation that longer design storm durations should be adopted for areas which are reservoired. The storm duration that is critical for a reservoir is longer than that required simply to generate peak flow off its catchment because of the greater storage volume involved. The report suggests a design storm duration for a reservoir equal to

(1 + average annual rainfall (mm)/1000)(T_p + reservoir lag)

where T_p is the time to peak of the unit hydrograph. It should be noted that

(a) lag refers to reservoir lag, i.e., the time between peak inflow and peak spillway outflow;
(b) iteration of calculations may be required because often reservoir lag can be estimated only by prior trial calculations;
(c) different flood control options can cause different peak outflow timings and therefore different design storm durations are required.
(d) T_p refers to the actual unit hydrograph used and thus design duration can shorten for the Probable Maximum Flood.

Table 3 Seasonal variation in Probable Maximum Precipitation*

Average annual rainfall, mm	Seasonal PMP value as percentage of all-year value†														
	Rainfall duration 1 h		Rainfall duration 2 h		Rainfall duration 6 h		Rainfall duration 1 day		Rainfall duration 2 days		Rainfall duration 4 days		Rainfall duration 6 days		Rainfall duration 25 days
	S‡	W‡	S	W	S	W	S	W	S	W	S	W	S	W	S	W
500–600	100	33	100	38	100	45	100	56	100	65	100	68	100	70	100	76
600–800	100	37	100	42	100	51	100	63	100	69	100	77	100	82	100	93
800–1000	100	47	100	50	100	61	100	72	100	80	100	84	100	91	100	99
1000–1400	100	63	100	69	100	79	100	84	100	88	100	92	100	97	96	100
1400–2000	100	74	100	86	100	93	100	99	97	100	95	100	92	100	90	100
Above 2000	100	82	100	90	100	96	97	100	90	100	91	100	89	100	80	100

*For winter rainfall in under 1 h, adopt the proportions of the 60 min value in Table II.3.6 of the *Flood studies report*.
†Snowmelt is not included.
‡Summer (S) is from May to October; winter (W) is from November to April.

CONSIDERATIONS WHEN USING THE FLOOD STUDIES REPORT

To summarize, design storm duration is contingent on the particular works being studied and, to a lesser extent, on the initial reservoir level adopted for routing studies.

Catchment subdivision. The *Flood studies report* method does not indicate that there need be any analysis other than that of the complete catchment to the dam site. However, it is advisable to combine floods from sub-areas of the catchment in certain circumstances. Examples are as follows.

If the reservoir area exceeds 5% of the direct catchment, rainfall onto the reservoir surface should be dealt with separately from that on the rest of the catchment. (This recognizes that rainfall arrives in the reservoir before tributary runoff.)

If the reservoir shape is such that it reaches well up the catchment, markedly reducing effective flow times to the dam site, the catchment should be divided into perhaps two or three stream areas draining to the reservoir perimeter. Alternatively, unit hydrograph response time for the 'lumped' catchment to the dam should be based on a typical stream draining to the reservoir edge rather than on the characteristics of the entire area.

If there are other reservoirs upstream of the one under consideration and these are capable of significant flood routing, analysis will be required over the sub-areas to each dam.

Seasonal variation of Probable Maximum Precipitation. Probable Maximum Precipitation (PMP) values can be obtained from the *Flood studies report* by a formal graphical factor method based on maps of 2 h and 24 h falls (Volume II §4.3.4); for durations of greater than 24 h Table II.4.4 should be used. The alternative method of multiplying up the five year return period fall should not be used. PMP obtained in this manner is an all-year limiting value; the Meteorological Office has approved a refinement of this calculation to estimate PMP values separately for summer and winter, so that only the latter need be compounded with snowmelt. The basis is the expectation that these seasonal values are in the same ratio as the 100 year values. Tables II.2.11 and II.3.9 from the *Flood studies report* are therefore combined and scaled to give Table 3. The all-year value of PMP for each duration of interest is assigned to the season shown as 100%, and PMP for the other season is scaled down from the all-year value by multiplying by the reduced percentage given. Table 3 does not immediately identify the season providing the design flood because snowmelt must be added to winter events.

Areal precipitation. Precipitation at a point is more easily obtained than the corresponding rainfall average over a given catchment area. It is doubtful whether reductions need be calculated over catchments of less than 10 km^2. The *Flood studies report* method for obtaining catchment area rainfall is in two stages, the first of which averages point values at several grid points within the area and the second of which adjusts this average point value to an empirically correct areal rainfall by a graph dependent on rain duration and total area. Such a method may be appropriate for stated probability events but has been criticized when applied to the concept of PMP. However, it is agreed that the procedure will give conservatively high results[21] and is therefore acceptable in the context of reservoir safety studies until further research offers improvement.

Precipitation over a cascade of reservoirs. It should be noted that the Meteorological Office offer a computer-based estimate of rainfall for any situation covered by the *Flood studies report*. As it is based on a data bank of results at a grid interval of 3.5 km it is most suited to studies of large catchments and areas where mapped rainfall gradients are small. It can simplify computation of concurrent rainfall over subcatchments within basins, as is needed for analysing a cascade of reservoirs. In the computation of PMP all the most severe rain intensities can be nested within one another in a single design event. Thus if PMP is computed for the reservoir most sensitive to a large volume over a long duration it will adequately test all the other reservoirs in that basin. However, appropriate allowance must be made for areal rainfall variations and this can be done by computing PMP separately for the total area to the lowest dam and for the one above it; the rainfall of the sub-area between the dams is then defined by the difference in volume in the two prior calculations. This procedure is repeated as necessary with any further dams higher up so that finally a consistent flood hydrograph can be routed down through the cascade.

Storm profile for large catchments. In catchments markedly exceeding 100 km^2, reservoir level can build up over a period of days as a consequence of a rapid succession of storms. In such cases it is inappropriate to assume a symmetrical storm profile of continuous rain. In a subject so little researched the best approach is to adopt the temporal pattern of the severest sequence of storms of the required duration that has been measured locally. The sequence with the most intense period at the end is the most severe case for a reservoir to withstand.

Local rainfall analyses. Local rainfall data should not normally be used to overrule the regional analysis performed by the Meteorological Office. This is because the length of additional record involved is usually so short by comparison with that needed for proper definition of rainfall distributions. There is evidence of under-estimates on *Flood studies report* rainfall maps for much of Somerset,[22] and for the few reservoirs affected it is suggested, because safety is involved, that local data should be adopted.

Local unit hydrograph data. Local data can be used to improve conversion of rainfall to an inflow hydrograph. If sufficient data are available a complete unit hydrograph should be derived, but first of all it should be ensured that the rainfall intensity is sufficiently high.

CONSIDERATIONS WHEN USING THE FLOOD STUDIES REPORT

Volume IV of the *Flood studies report* details results with which comparisons can be made. If information is more limited it may be possible to measure the time lag from the centroid of rainfall to the peak of the ensuing flood in several major floods; T_p is then estimated as $0.9 \times \text{lag}$.

Local percentage runoff data. Local figures can be sufficiently convincing to revise percentage runoff. This factor is mainly controlled by soil type and the country is classified into only five types for *Flood studies report* purposes. There is some evidence for runoff proportions above Type 5[23,24] from hard rock and permanently saturated soils in mountainous areas and if local data confirms this it should be adopted in reservoir flood studies.

Computation time steps. Rainfall increments should be of sufficiently small duration to match the subsequent flood routing procedure. Computer studies will be based normally on time steps rounded down from the definition of $0.2T_p$ in the *Flood studies report*. Units of 30 min, 20 min or 15 min suit typical UK reservoirs. It is preferable for design storm duration to be divided by an odd integer number of these time steps so that a symmetrical rain storm is easily organized; 20 min steps readily produce this.

Variation of RSMD with area. A check on the computation of RSMD, the *Flood studies report* index for flood-producing rainfall, is provided by Figs 5–7. These figures show point values; reductions for large catchments can be made.[25] A reduction of 3% is suggested for 10 km², increasing to 7% over 100 km².

Antecedent flow. Certain categories in Table 1 require average flow to be passing down the spillway at the start of the flood. This flow, the long term mean inflow from the direct catchment, is intended to correspond to the baseflow of the final step outlined in section 3.2. The *Flood studies report* suggests a complex antecedent flow (ANSF) defined by a formula drawn mainly from large catchment floods; it gives a value near to winter average flow for a typical upland catchment. However, for simplicity the better known value of mean inflow is preferred here. On the comparatively few dam catchments exceeding 100 km² a local study is warranted to check the consequences of adopting the higher value of ANSF. For floods extending over several days in a large basin this antecedent flow should be allowed to decay with time to represent discharge by ground wet from previous rainfall. If left at a constant value, too high a flood volume will be obtained.

Glossary

Conformity with the ICOLD *Glossary of words and phrases related to dams*[26] has not been found practical; the glossary which follows should only be used in the context of this guide as some definitions have been deliberately restricted.

Auxiliary spillway	A secondary spillway designed to operate only during exceptionally large floods[26]
Emergency spillway	As Auxiliary spillway
Fetch	The effective reservoir length over which winds can build up waves
Flood surcharge	The maximum rise of still water level above reservoir top water level during a design flood. (Surcharge water is not retained in the reservoir but is discharged until the normal retention level is reached.)
Freeboard, Dam	The vertical height from top water level to the top of the dam. (Freeboard is required to contain flood surcharge plus wave surcharge of specified severity.)
Freeboard, Wave	The vertical height remaining between the top of the dam and the height reached by the flood surcharge
Mean annual flood	The arithmetic mean of an annual maximum series of instantaneous flood peaks
Normal retention level	As Top water level
Probable Maximum Precipitation (PMP)	'The (theoretical) greatest depth of precipitation for a given duration meteorologically possible for a given basin at a particular time of year...'[27] It includes rain, sleet, snow and hail as it occurs, but not snow cover left from previous storms

Probable Maximum Flood (PMF)	The flood hydrograph resulting from PMP and, where applicable, snowmelt, coupled with the worst flood-producing catchment conditions that can be realistically expected in the prevailing meteorological conditions
RSMD	The *Flood studies report* index for flood-producing rainfall defined as the one day rainfall of five year return period less effective mean soil moisture deficit
Rainfall excess	Precipitation minus losses; i.e., equivalent to flood runoff volume over and above baseflow
Reservoir flood routing	The passage of a flood wave through a reservoir. Sometimes used to describe the calculation of the attenuation of the hydrograph of the incoming flood as it passes through storage and down the spillway
Return period	The average expected time (in terms of probability rather than forecasting) between floods equal to or greater than a stated magnitude
Rip-rap	Graded quarry stone, usually placed on a graded filter
Run-up	The maximum vertical height attained by a wave running up a dam face, referred to the steady water level without wind action
Significant wave height	The wave height, trough to crest, that is exceeded by only a small stated percentage of waves
Storm losses	That part of precipitation which does not become runoff within the flood period because it has evaporated, infiltrated, been retained in the soil or been temporarily ponded on the catchment surface
Storm profile	The magnitude and sequence of precipitation in equal time increments during a storm of given duration

GLOSSARY

Top water level	(*a*) For a reservoir with a fixed overflow sill, the lowest crest level of that sill; (*b*) for a reservoir from which the overflow is controlled wholly or partly by movable gates, siphons or other means, the maximum level to which water may be stored exclusive of any provision for flood storage. At this level the reservoir is 'just full' (see Table 1)
Unit hydrograph (UH)	The runoff hydrograph resulting from unit volume of rainfall excess in a specified duration of time over a given catchment; the rainfall is presumed to fall uniformly or characteristically in both time and space on the catchment in the specified duration
Wave, Dam break	The surge wave of water released down valley by a dam that fails
Wave surcharge	The rise of water against a dam created solely by the run-up of waves of specified probability
Wave surcharge allowance	For the particular type and design of dam under consideration, the theoretical wave freeboard sufficient to prevent overtopping reaching quantities that could threaten the dam crest

References

1. NATURAL ENVIRONMENT RESEARCH COUNCIL. *Flood studies report.* NERC, London 1975. (Available from the Institute of Hydrology, Wallingford.)
2. INSTITUTION OF CIVIL ENGINEERS, FLOODS WORKING PARTY. *Discussion paper on reservoir flood standards.* Institution of Civil Engineers, London, 1975.
3. AUSTRALIAN NATIONAL COMMITTEE ON LARGE DAMS. *Guidelines for operation, maintenance and surveillance of dams.* Australian National Committee on Large Dams, Melbourne, 1976.
4. BINNIE G. M. et al. Discussion on reservoir flood standards. *Flood studies conference.* Institution of Civil Engineers, London, 1975, 87–106.
5. LAW F. M. et al. (MOFFAT A. I. B. (ed.)). Session on flood analysis, prediction and design in relation to reservoirs, dams and spillways. *Inspection, operation and improvement of existing dams.* University of Newcastle, 1975. Proceedings of BNCOLD/University of Newcastle symposium.
6. INSTITUTE OF HYDROLOGY. *Guide to the flood studies report.* Institute of Hydrology, Wallingford, 1978.
7. KIRKPATRICK G. W. Guidelines for evaluating spillway capacity. *Int. Wat. Pwr Dam Constr.*, 1977, **29**, Aug., No. 8, 31, Fig. 4.
8. McQUIVEY R. S. and KEEFER T. N. Application of simple dam break routing model. *Proc. Int. Assn Hydraulic Research 16th congress, Sao Paulo, 1975.* Organizing Committee of the Congress, Sao Paulo, **2**, 315–324.
9. AMERICAN SOCIETY OF CIVIL ENGINEERS, COMMITTEE ON HYDROMETEOROLOGY, TASK COMMITTEE. Reevaluating spillway adequacy of existing dams. *J. Hydraul. Div. Am. Soc. Civ. Engrs*, 1973, **99**, Feb., HY2, 337–372.
10. GRIFFITHS F. N. and BERRY D. W. (MOFFAT A. I. B. (ed.)). Spillways — design philosophy. *Inspection, operation and improvement of existing dams.* University of Newcastle, 1975. Proceedings of BNCOLD/University of Newcastle symposium.
11. CONSTRUCTION INDUSTRY RESEARCH AND INFORMATION ASSOCIATION. *A guide to the use of grass in hydraulic engineering practice.* CIRIA, London, 1976, Tech. note 71, 16.

12. TWORT A. C. et al. *Water supply.* Edward Arnold, London, 1974, 2nd edn, 61–63.
13. NATURAL ENVIRONMENT RESEARCH COUNCIL. *Flood studies report.* NERC, London, 1975, 1, chapter 7 (including a Fortran IV program listing).
14. COLOMBI J. S. and HALL M. J. A quick screening method for estimating the routeing effect of a reservoir. *Proc. Instn Civ. Engrs,* Part 2, 1977, 63, Dec., 935–941.
15. JARVIS R. M. Flood assessment for hydro-electric projects. *Inspection, operation and improvement of existing dams.* University of Newcastle, 1975. Proceedings of BNCOLD/ University of Newcastle symposium.
16. BRITISH HYDROMECHANICS RESEARCH ASSOCIATION. *Proceedings of symposium on design and operation of siphons and siphon spillways.* Cranfield BHRA Fluid Engineering, Cranfield, 1976.
17. SAVILLE T. et al. Freeboard allowances for waves in inland reservoirs. *J. WatWays Harb. Div. Am. Soc. Civ. Engrs,* 1962, 88, May, WW2, 102.
18. TECHNICAL ADVISORY COMMISSION ON PROTECTION AGAINST INUNDATION. *Wave run-up and overtopping.* Netherlands Government Publishing Office, The Hague, 1974, 16.
19. COCHRANE N. J. An engineering calculation of risk in the provision for the passage of floods during the construction of dams. *Trans 9th congress International Commission on Large Dams, Istanbul, 1967,* 5, 325–341.
20. GOLDING D. Googong Dam's 100-year flood. *J. Instn Engrs Aust.,* 1976, 48, 5 Nov., 16.
21. BELL F. C. *The areal reduction factor in rainfall frequency estimation.* Institute of Hydrology, Wallingford, 1976, Report No. 35.
22. BOOTMAN A. P. and WILLIS A. *Extreme two day rainfall in Somerset.* Wessex Water Authority, Bristol, 1977.
23. REYNOLDS G. (MOFFAT A. I. B. (ed.)). Extreme rainfall estimation for flood studies in the Scottish Highlands. *Inspection, operation and improvement of existing dams.* University of Newcastle, 1975. Proceedings of BNCOLD/University of Newcastle symposium, 14.4–4.
24. INSTITUTE OF HYDROLOGY. *Report of the seminar 'The flood studies report — an opportunity for discussion'.* Institute of Hydrology, Wallingford, 1977. Flood studies supplementary report No. 3.
25. FARQUHARSON F. A. K. et al (MOFFAT A. I. B. (ed.)). Some aspects of design flood estimation. *Inspection, operation and improvement of existing dams.* University of Newcastle, 1975. Proceedings of BNCOLD/University of Newcastle symposium.
26. INTERNATIONAL COMMISSION ON LARGE DAMS, COMMITTEE ON THE DICTIONARY, THE GLOSSARY

REFERENCES

AND THE WORLD REGISTER OF DAMS. *Glossary of words and phrases related to dams.* ICOLD, Paris, 1977, Bulletin 31.

27. WORLD METEOROLOGICAL ORGANISATION. *Manual for estimation of probable maximum precipitation.* WMO, Geneva, 1973. Operational hydrology report No. 1, (xiii).

Bibliography

General

AMERICAN SOCIETY OF CIVIL ENGINEERS. *Inspection, maintenance and rehabilitation of old dams.* ASCE, New York, 1974.

BENNET G. Bristol floods 1968: controlled survey of effects on health of local community disaster. *Br. Med. J.*, 1970, **3**, 22 Aug., 454–458.

DEPARTMENT OF THE ENVIRONMENT. *Reservoirs Act 1975: consultative document concerning draft forms.* Private communication, 1976.

GERMOND J. P. Insuring dam risks. *Int. Wat. Pwr Dam Constr.*, 1977, **29**, June, No. 6, 36–39.

INSTITUTION OF CIVIL ENGINEERS. *Reservoir safety: report of the ad hoc Committee to submit proposals for a revision of the Reservoirs (Safety Provisions) Act 1930.* ICE, London, 1966.

INTERNATIONAL COMMISSION ON LARGE DAMS. *Report of Committee on Risks to Third Parties from Large Dams.* ICOLD, Paris, 1977.

INTERNATIONAL COMMISSION ON LARGE DAMS, US COMMITTEE. *Criteria and practices utilized in determining the required capacity of spillways.* USCOLD, New York, 1970.

POCHIN E. E. The acceptance of risk. *Br. Med. Bull.*, 1975, **31**, Sept., No. 3, 184–190.

SECRETARY OF STATE et al. *Statutory rules and orders, 1930, No. 1125: reservoirs – regulations, Reservoirs (Safety Provisions) Act 1930.* HMSO, London, 1969.

TITCHENER J. L. and KAPP F. T. Family and character change at Buffalo Creek. *Am. J. Psychiat.*, 1976, **133**, Mar., No. 3, 295–299.

Technical

BUTTERS K. and LANE J. J. Flood alleviation works on the River Thames tributaries. *J. Instn Wat. Engrs*, 1975, **29**, Mar., No. 2, 67–94.

DAVIDSON D. D. and McCARTNEY B. L. Water waves generated by landslides in reservoirs. *J. Hydraul. Div. Am. Soc. Civ. Engrs*, 1975, **101**, Dec., HY12, 1489–1501.

INSTITUTION OF CIVIL ENGINEERS. *An introduction to engineering economics.* ICE, London, 1969, 45–50, Example 5.3.

INTERNATIONAL COMMISSION ON LARGE DAMS. *Trans. 11th congress ICOLD, Madrid, 1973.* **III:** Slope protection.
MILLER D. L. (US BUREAU OF RECLAMATION (ed.)). Inflow design flood studies. *Design of gravity dams.* US Government Printing Office, Denver, Colorado, 1976, 435–509.
OLIVER G. C. S. Eyebrook Reservoir siphon spillway—paper II. *J. Instn Wat. Engrs,* 1976, **30,** Feb., No. 1, 35–42.
PENNING-ROWSELL E. C. and CHATTERTON J. B. *The benefits of flood alleviation: a manual of assessment techniques.* Saxon House, Farnborough, 1977.

FIGURES

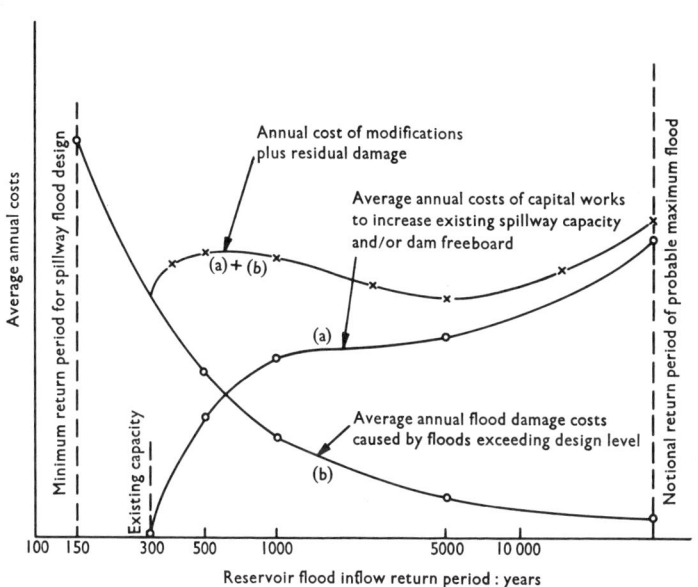

Fig. 1. Least total cost analysis

FLOODS AND RESERVOIR SAFETY

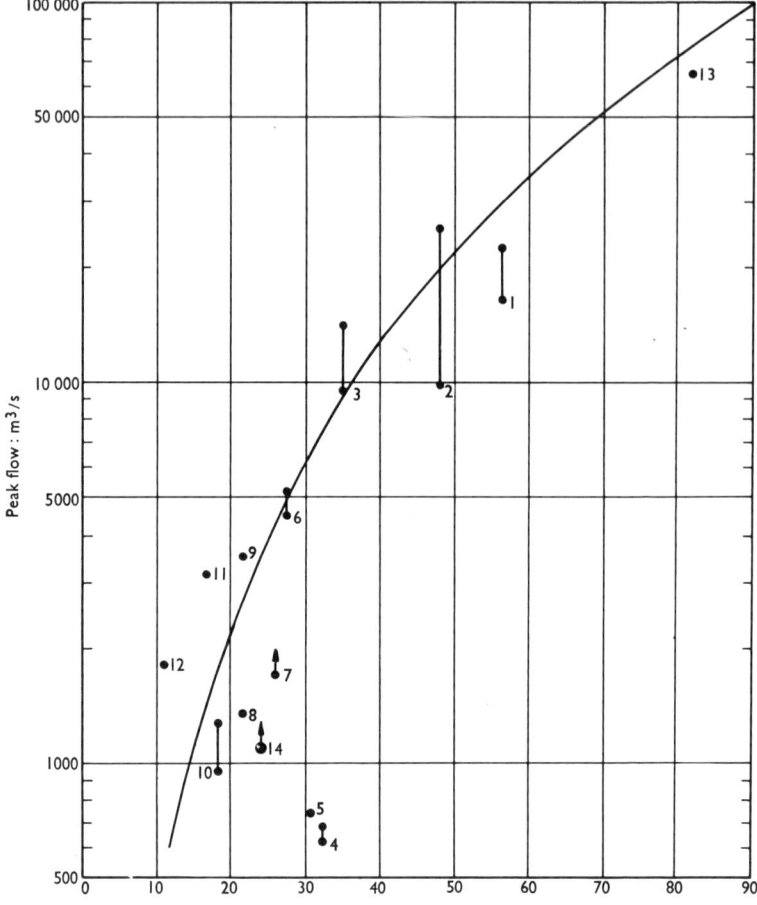

Dam height if dam overtopped, or depth of water at time of failure if dam not overtopped; H: m

Estimated flood peaks from dam failures. The numbers indicate the name of the dam, its location, type of dam where known, and the year of failure.
1. St Francis, California, concrete gravity, 1928
2. Swift, Montana, rock fill, 1960
3. Oros, Brazil, earth and rock fill, 1960
4. Apishapa, Colorado, earth fill, 1923
5. Hell Hole, California, rock fill, 1964
6. Schaeffer, Colorado, earth fill, 1921
7. Granite Creek, Alaska, 1971, discharge at 8 km downstream
8. Little Deer Creek, Utah, earth fill, 1963
9. Castlewood, Colorado, rock fill, 1933
10. Baldwin Hills, California, earth fill, 1963
11. Hatchwood, Utah, earth fill, 1914
12. Lower Two Medicine, Montana, 1964
13. Teton dam, Idaho, earth, 1976
14. Dale Dyke, Sheffield, England, earth fill, 1864, discharge at 10 km downstream

Fig. 2. Dam failure flood flow v. dam height (after Kirkpatrick[7])

FIGURES

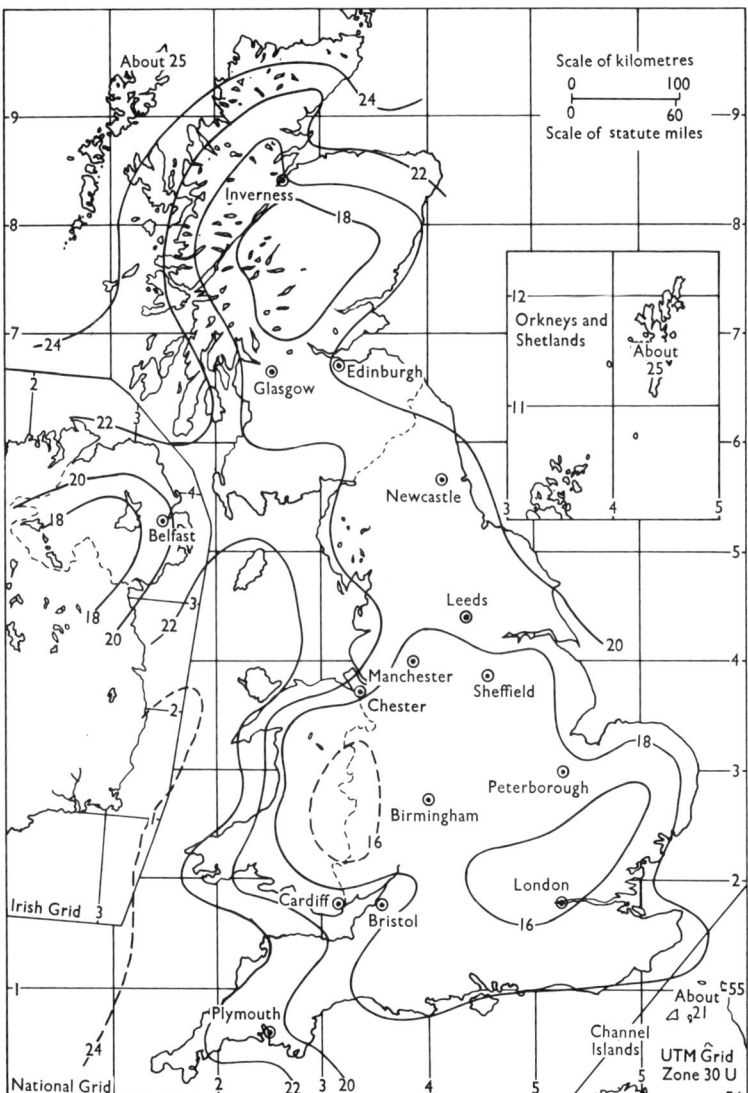

Fig. 3. Average hourly annual maximum wind speed (m/s)

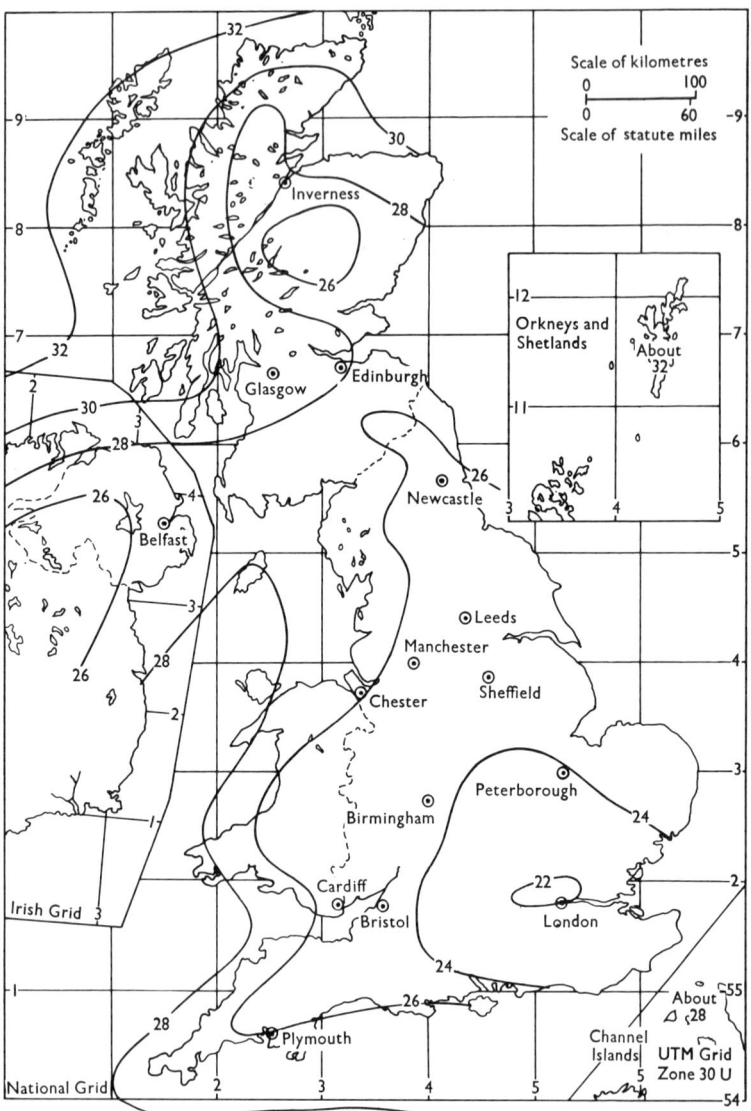

Fig. 4. Once in ten year hourly maximum wind speed (m/s)

FIGURES

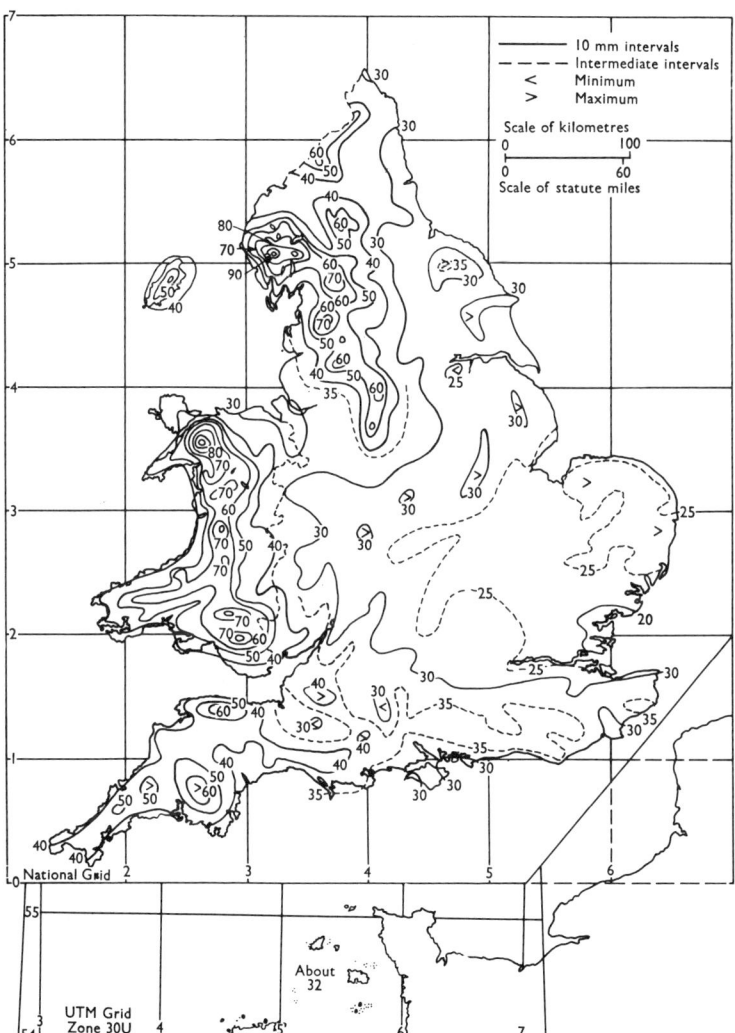

Fig. 5. RSMD (mm): England and Wales

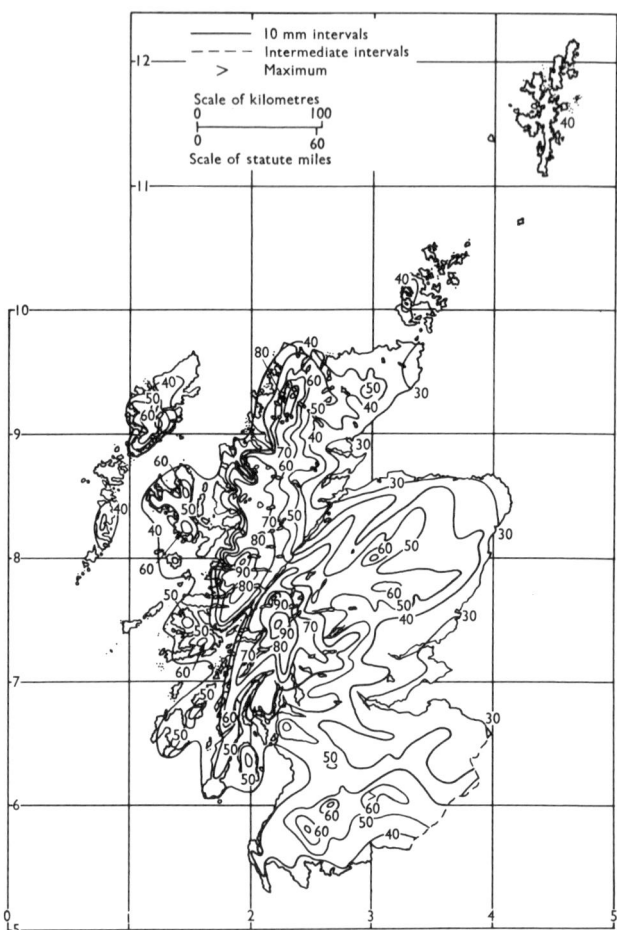

Fig. 6. RSMD (mm): Scotland

FIGURES

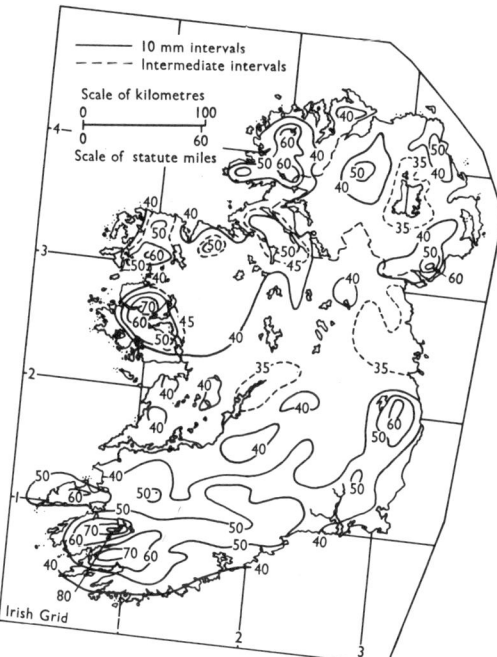

Fig. 7. RSMD (mm): Ireland

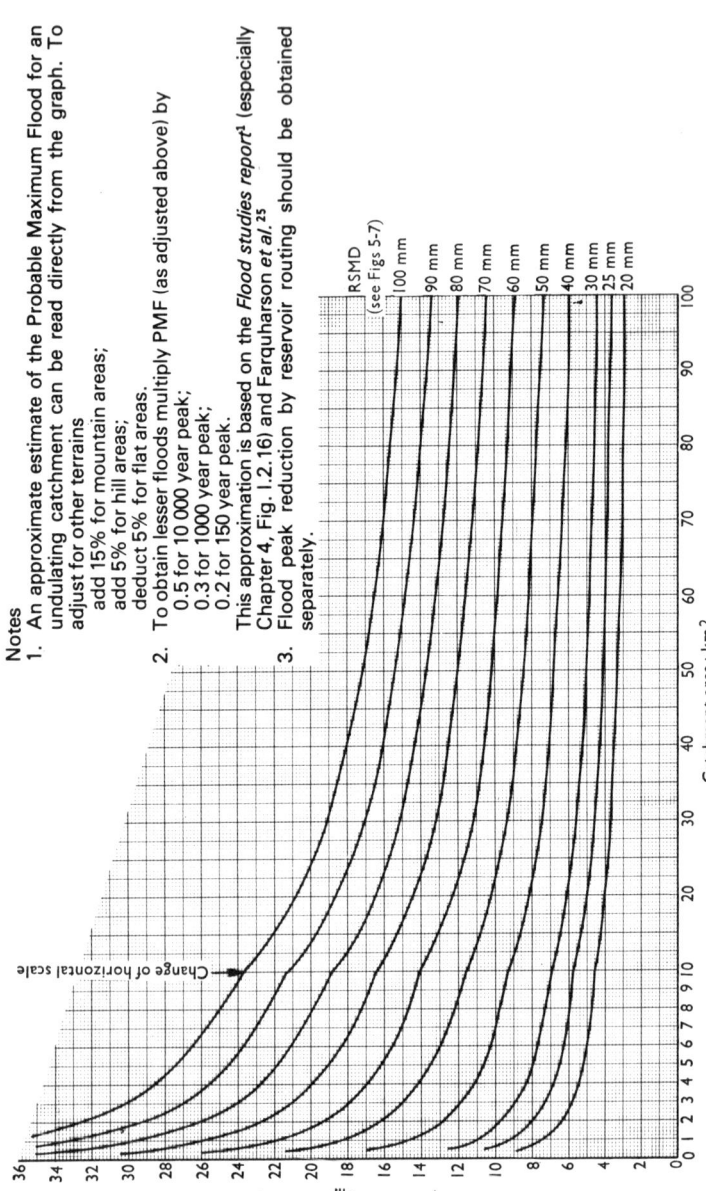

Fig. 8. Flood peak intensity for impermeable reservoired catchments: curves based on equation (2) of Farquharson *et al.*[25]

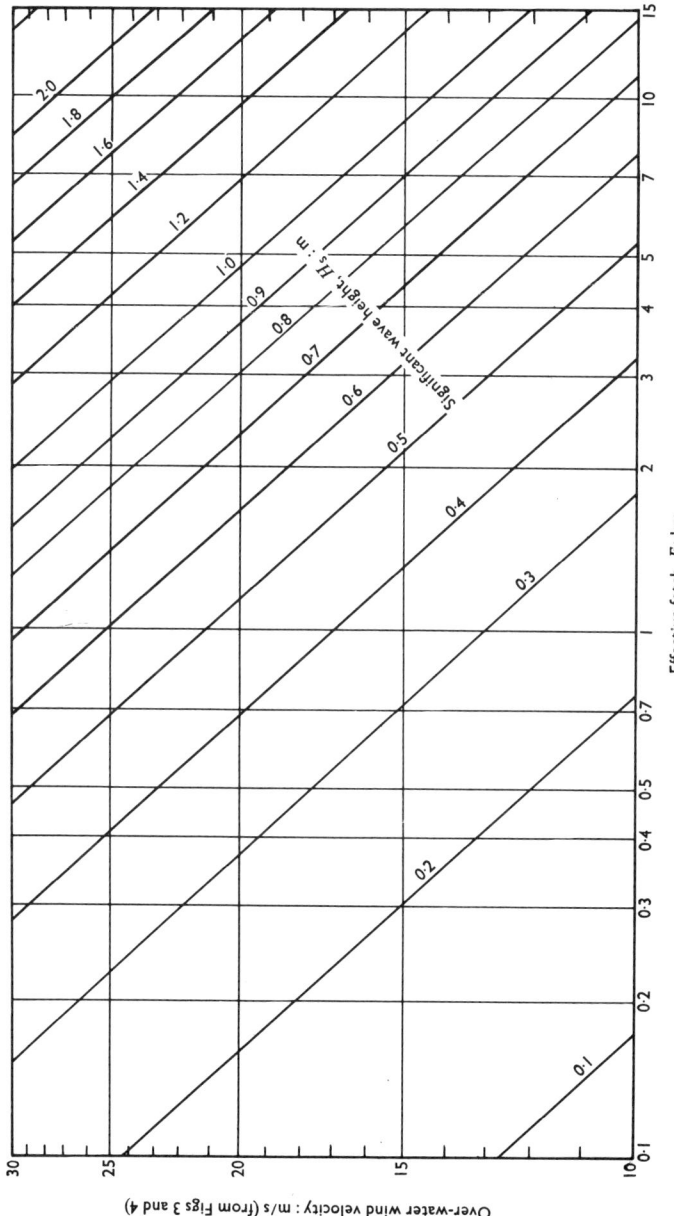

Fig. 9. Relationship between effective fetch, wind speed and significant wave height

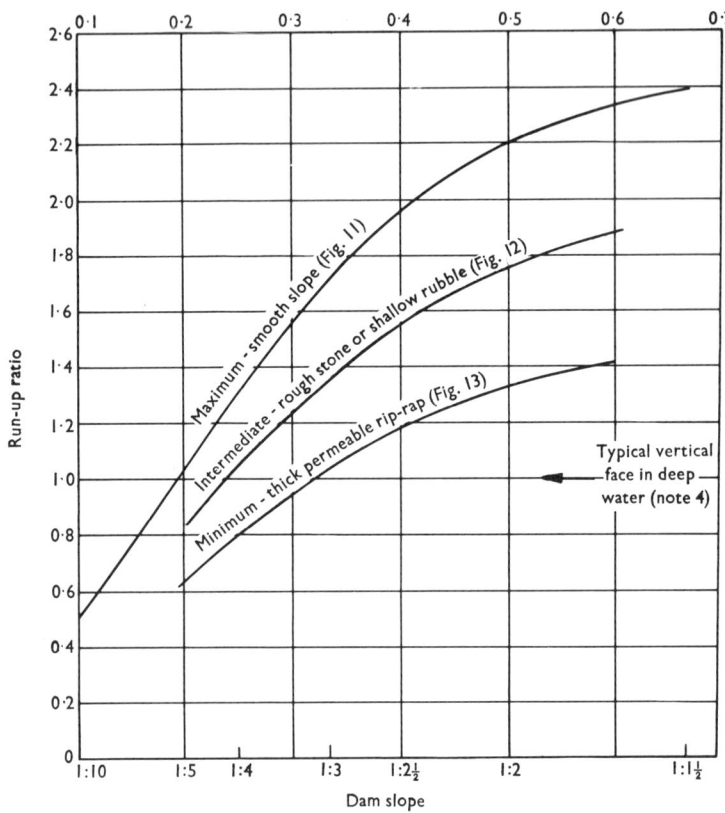

Notes
1. Maximum line from Saville et al.[17] p. 115 for typical wave steepness (height/length) = 0.05.
2. Intermediate line is 0.8 × maximum from Technical Advisory Commission on Protection Against Inundation[18] p. 69.
3. Minimum line is 0.6 × maximum from Technical Advisory Commission on Protection Against Inundation[18] pp. 69 and 133.
4. For faces off-vertical the run-up ratio rises above unity and can approach 2 in some circumstances where the deep water condition is not fulfilled.

Fig. 10. Ratio of run-up height to design wave height

FIGURES

Fig. 11. Typical smooth dam facing: maximum run-up (Grafham Water dam, Huntingdon; Anglian Water Authority)

Fig. 12. Rough open-jointed pitching: intermediate run-up (Todd Brook dam (pre 1969); British Waterways Board)

Fig. 13. Thick rip-rap facing: minimum run-up (Draycote Water; Severn–Trent Water Authority)

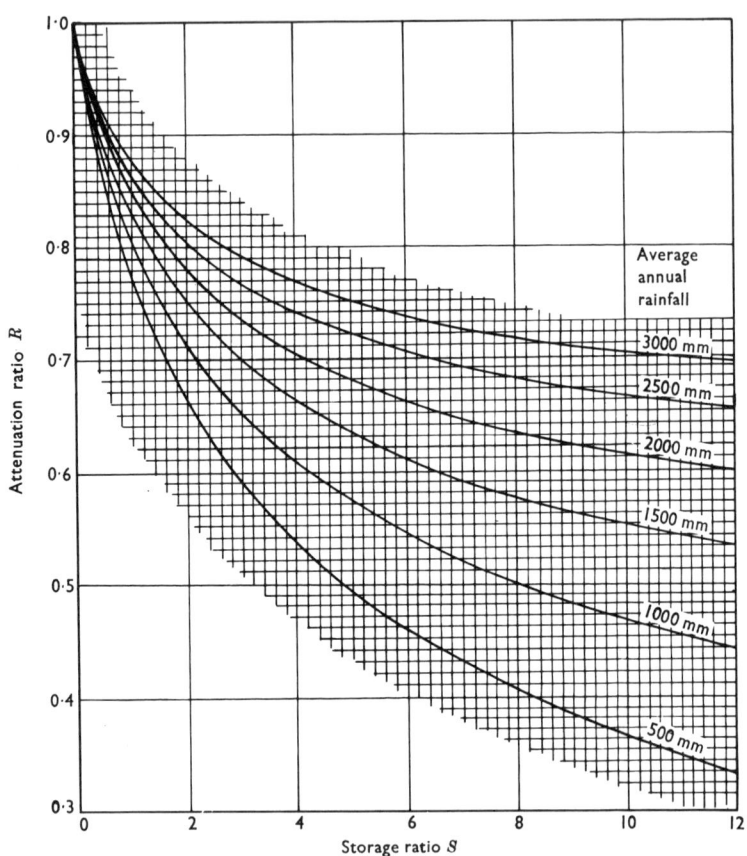

Fig. 14. Flood routing graph (after Colombi and Hall[14])